Habitats

by Fran Downey

Contents

Introduction: Our Planet 4

Chapter 1

Big Idea Question

Where Do Plants and Animals Live? 6
    Water Habitats 8
    Land Habitats 10
    A Forest Habitat 12

Chapter 2

Big Idea Question

What Do Plants and Animals Need to Survive? 16
    Survival 18

Chapter 3

Big Idea Question

How Do Plants and Animals Depend on Each Other? ... 24
    Animals Need Plants 26
    Plants Need Animals 32

Conclusion: Life on Planet Earth 36

Glossary 38
Index 40

Introduction

Our Planet

Next Generation Sunshine State Standards
SC.2.L.17.2 Recognize and explain that living things are found all over Earth, but each is only able to live in habitats that meet its basic needs.

If you could zoom into space and look back at **Earth**, you would see blue water, green and brown land, and white clouds. Beautiful, isn't it? And it's a great place to live!

Blue water

Green and brown land

White clouds

Chapter 1

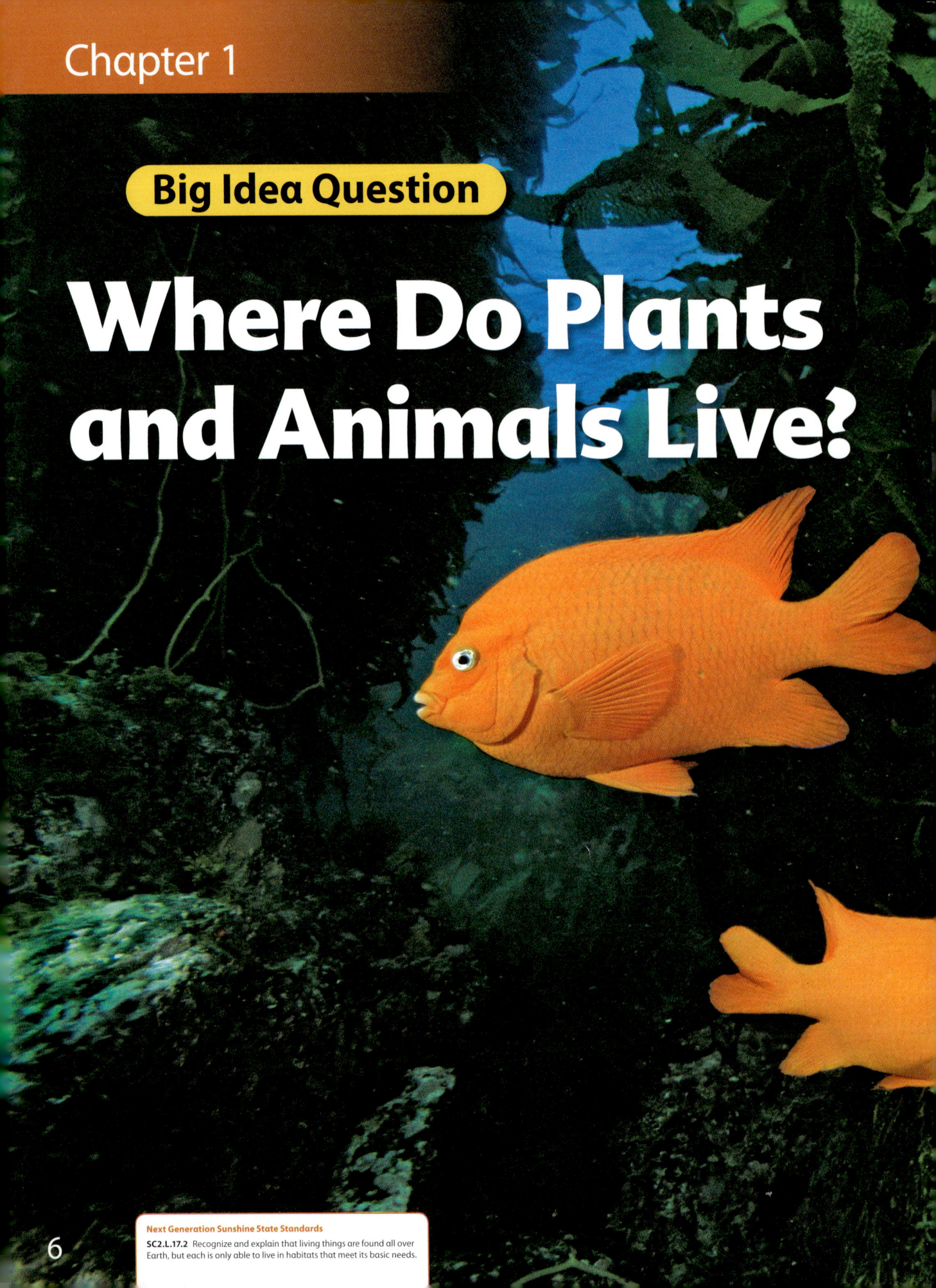

Big Idea Question

Where Do Plants and Animals Live?

Next Generation Sunshine State Standards
SC2.L.17.2 Recognize and explain that living things are found all over Earth, but each is only able to live in habitats that meet its basic needs.

Plants and animals live almost every place on Earth. These places can be wet or dry or hot or cold.

A place where plants and animals get what they need to stay alive is a **habitat**. What are Earth's habitats like? What kinds of plants and animals live in these habitats? Let's find out.

Water Habitats

Most of Earth is covered by water. Some plants and animals live in the saltwater ocean.

These sea lions dive into the ocean.

Other kinds of plants and animals live in fresh water. Some live in a quiet pond. Some live in a rushing river.

Pond

These geese are one kind of living thing in this pond.

River

The fish are swimming upstream.

Land Habitats

Some plants and animals can never live in water. They live and grow on land.

These plants can grow in a desert.

There are different kinds of land habitats. Prairies, forests, and deserts are a few examples. How are they different?

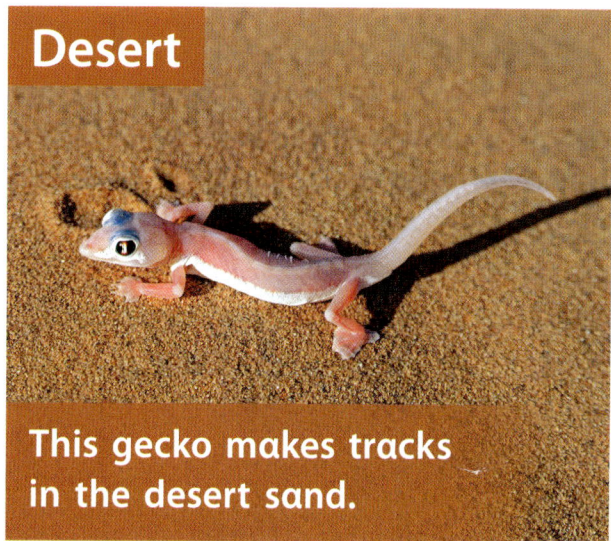

Prairie

These bison graze on a North American prairie.

Forest

This wolf looks for food in a forest.

Desert

This gecko makes tracks in the desert sand.

A Forest Habitat

Let's explore a forest habitat. In many forests, you would find lots of different trees. Some trees have wide leaves. Others have needles. A needle is a kind of narrow leaf.

Trees with leaves

Wide leaf

Trees with needles

Needles

Many animals live in a forest habitat. In some forests, squirrels build leaf nests in trees. Woodpeckers find insects to eat in the trees. What other animals might live in a forest habitat?

A moose finds grass to eat in a forest.

Forests are habitats for many animals, including deer, salamanders, woodpeckers, black bears, and red foxes. A forest is one place where many plants and animals live.

This deer lives in a forest.

This cave salamander lives in a forest cave.

A woodpecker finds insects to eat.

A black bear finds a fish to eat.

A red fox finds a place to hide.

Chapter 2

Big Idea Question

What Do Plants and Animals Need to Survive?

Next Generation Sunshine State Standards
SC.2.L.17.1 Compare and contrast the basic needs that all living things, including humans, have for survival.
SC.2.L.17.2 Recognize and explain that living things are found all over Earth, but each is only able to live in habitats that meet its basic needs.

A habitat is important. It provides plants and animals with air, water, food, and space—all the things they need to stay alive, or **survive**.

But a habitat has to be the right match for each animal. For example, a squirrel could not survive in the ocean. And a whale could not survive in a forest.

Survival

All animals have basic needs. They must have food, air, and water. They must have space or **shelter**. A shelter can be a home or safe place.

Food

A western bluebird finds food to eat.

Water

A wolf finds water to drink.

Shelter

These raccoon babies are in a shelter.

All plants have basic needs, too. They must have light, water, air, and space to live and grow in their habitats.

Trees use sunlight to make their own food.

Plants and animals have special parts, such as a bird's feathers. Most birds use their feathers and wings to fly. These parts help them survive.

A beaver's sharp teeth help it cut twigs and branches.

A duck's feathers help it fly.

The colors of some wildflowers attract bees.

This owl's good eyesight helps it hunt for food.

A deer has white and tan fur. The two colors of fur help hide the deer. It blends in with the forest.

The young deer, or fawn, blends in with the sunlight and shadows of the forest. This helps it survive.

Chapter 3

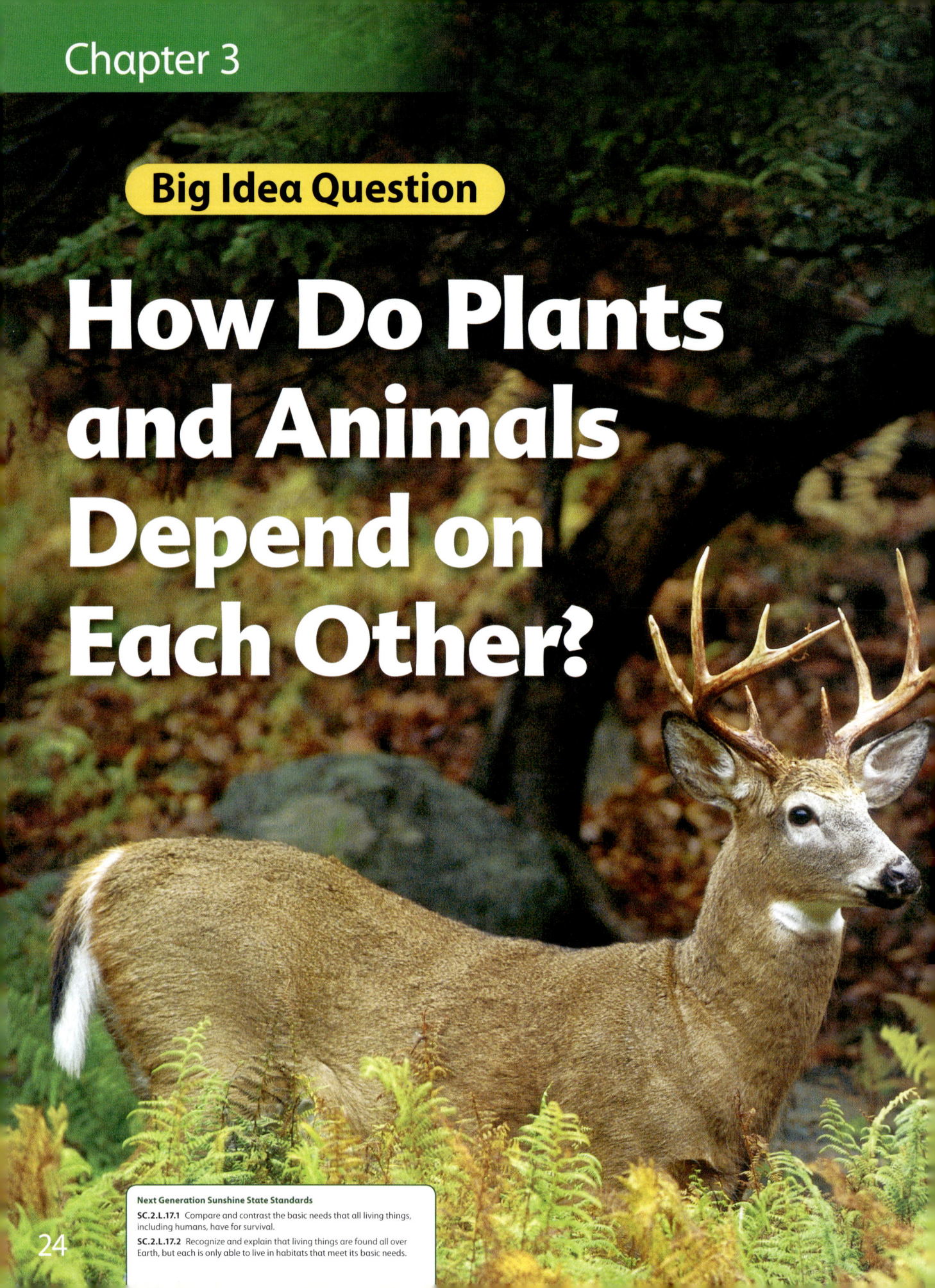

Big Idea Question

How Do Plants and Animals Depend on Each Other?

Next Generation Sunshine State Standards
SC.2.L.17.1 Compare and contrast the basic needs that all living things, including humans, have for survival.
SC.2.L.17.2 Recognize and explain that living things are found all over Earth, but each is only able to live in habitats that meet its basic needs.

Can you imagine a world without plants? Animals could not survive without plants. But some plants could survive without animals.

Animals Need Plants

Plants make their own food, using sunlight, air, water, and **nutrients**. Nutrients are parts of food and soil that help plants and animals grow.

This apple tree makes its own food.

This rabbit eats plants.

Animals can't make their own food. Some animals eat only plants. Some eat only animals. And some animals eat both plants and animals.

This cougar eats other animals.

This coyote eats both plants and animals.

Plants and animals depend on each other. Plants get **energy** from the sun. Energy is the ability to do active things.

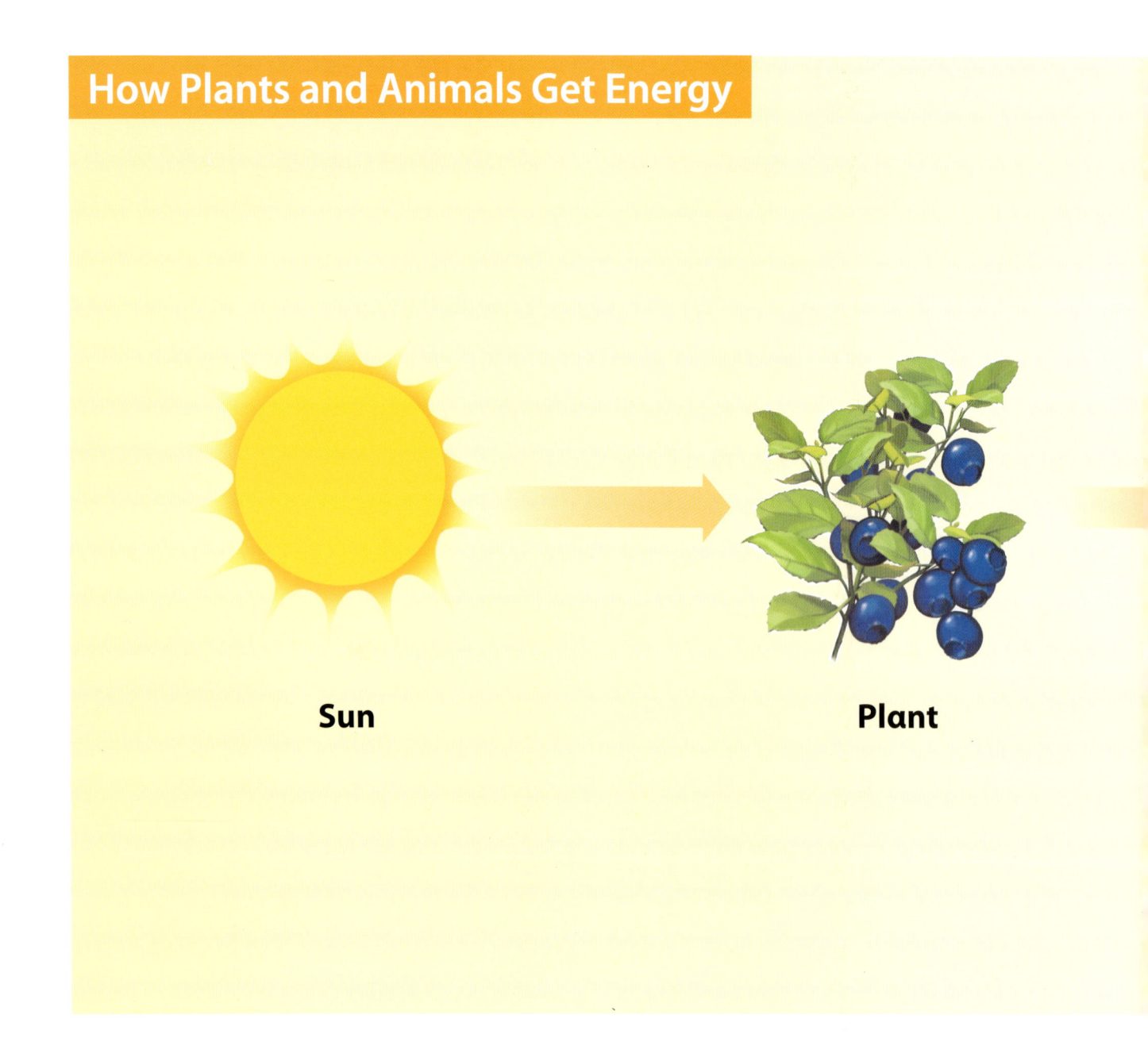

How Plants and Animals Get Energy

Sun

Plant

A mouse eats plants for food. It gets energy from the plant. When an owl eats the mouse for food, it gets energy from the mouse.

Mouse

Owl

Animals need plants for more than food. Plants give off **oxygen**. It is the part of air that animals breathe in to survive.

This tree gives off oxygen.

This elk breathes in oxygen.

So animals really do depend on plants. They need plants for oxygen. They need plants for food. They also need plants for shelter.

This hummingbird gets nectar from flowers.

This bird uses grass to make its nest.

Plants Need Animals

Why do some plants need animals? Most plants are rooted in the ground. So many plants need animals to carry their seeds. This way, plants can grow in new places.

This squirrel is moving an acorn.

This nuthatch is carrying a seed.

There is another way that animals help plants. Animals dig in the soil. This loosens the soil and makes it easier for plants to grow there.

This badger is loosening the soil.

Animal waste also helps enrich soil. Dead and decaying plants and animals do, too. They add nutrients back to the soil.

Nutrients enrich the soil, which helps new plants grow.

Conclusion

Life on Planet Earth

Earth has many habitats. Some are water habitats. Others are land habitats. Plants and animals live in these habitats.

Plants and animals get what they need to survive in their habitats. Their special parts help them get what they need.

All plants and animals in a habitat help one another survive.

37

Glossary

Earth (page 5)
Earth is the planet on which we live.

Earth has many habitats.

energy (page 28)
Energy is the ability to do active things.

A mouse gets **energy** by eating plants.

habitat (page 7)
A **habitat** is a place where living things can get what they need to stay alive.

Some fish live in an ocean **habitat.**

nutrients (page 26)
Nutrients are parts of food and soil. They help living things stay healthy and grow.

Animals get **nutrients** from the food they eat.